PLANTS with SUPERPOWERS

PLANTS CAN MIMIC!

Monika Davies

Gareth Stevens PUBLISHING

Please visit our website, www.garethstevens.com. For a free color catalog of all our high-quality books, call toll free 1-800-542-2595 or fax 1-877-542-2596.

Cataloging-in-Publication Data
Names: Davies, Monika.
Title: Plants can mimic! / Monika Davies.
Description: New York : Gareth Stevens Publishing, 2023. | Series: Plants with superpowers | Includes glossary and index.
Identifiers: ISBN 9781538278956 (pbk.) | ISBN 9781538278970 (library bound) | ISBN 9781538278963 (6pack) | ISBN 9781538278987 (ebook)
Subjects: LCSH: Plant defenses–Juvenile literature. | Mimicry (Biology)–Juvenile literature. | Plants–Adaptation–Juvenile literature.
Classification: LCC QK921.D38 2023 | DDC 581.4'7–dc23

Published in 2023 by
Gareth Stevens Publishing
29 East 21st Street
New York, NY 10010

Editor: Monika Davies
Designer: Leslie Taylor

Portions of this work were originally authored by Louella Bath and published as *Plants that Mimic*. All new material in this edition authored by Monika Davies.

Photo credits: Cover (photo) Grazyna Palaszewska/Shutterstock.com; Cover (font) Alhovik/Shutterstock.com; Series Art (various accents) Ozz Design, Piotr Urakau, Designer things, Maxger/Shutterstock.com; p. 5 ArgenLant/Shutterstock.com; p. 7 (fly) Cathy Keifer/Shutterstock.com; p. 7 (sundew) W. de Vries/Shutterstock.com; p. 9 (upper left) shutting/Shutterstock.com; p. 9 (lower right) AjayTvm/Shutterstock.com; p.9 (lower left) kunanon/Shutterstock.com; p. 11 (living stones) SIM ONE/Shutterstock.com; p. 11 (bloom) schlyx/Shutterstock.com; p. 13 (bee) Viktor Loki/Shutterstock.com; p. 13 (bee orchid) PJ photography/Shutterstock.com; p. 15 (fly) De19/Shutterstock.com; p. 15 (fly orchid) Douwe Schut/Shutterstock.com; p. 17 (left) https://commons.wikimedia.org/wiki/File:Boquila_trifoliata.jpg; p. 17 (right) https://commons.wikimedia.org/wiki/File:Boquila_trifoliolata_3.jpg; p. 19 (upper left) Gerry Bishop/Shutterstock.com; p. 19 (lower right) NatureDiver/Shutterstock.com; p. 19 (fly) Tomasz Klejdysz/Shutterstock.com; p. 21 (upper left) Darren Kurnia/Shutterstock.com; p. 21 (right) Erik Cox Photography/Shutterstock.com.

Printed in the United States of America

CPSIA compliance information: Batch #CSGS23: For further information, contact Gareth Stevens, New York, New York, at 1-800-542-2595.

CONTENTS

Words in the glossary appear in **bold** type the first time they arew used in the text.

Hidden in Plain Sight

Every animal and plant wants to live a long life. Plants come up with all kinds of **sneaky** tricks to stay alive. In fact, some plants have learned how to trick the eyes of bugs and other animals. These plants can hide in plain sight—or draw in certain bugs—using mimicry!

Mimicry is when a plant looks or smells like something else, such as another animal or plant. Plants that mimic are masters of **disguise**. Let's learn more about these tricky plants.

Super Fact!

The *Lamium album* plant is nicknamed "dead nettle." That's because it looks a lot like the stinging nettle plant! The stinging nettle has stinging hairs, and most animals avoid it. However, the dead nettle lacks the same harmful sting.

By copying the stinging nettle's looks, the dead nettle stays safe from animals that might eat it.

Sneaky Adaptations

Every **habitat** presents challenges for the creatures living there. There could be a lack of sunlight, cold weather, or nasty predators. Over thousands of years, plants have developed ways of dealing with these problems.

Mimicry is a kind of adaptation. An adaptation is a change in a living thing that makes it better able to live in its surroundings. Years ago, plants that displayed mimicry survived longer than plants that didn't. These plants passed on their **traits** to new plants, and the traits became part of the species.

Super Fact!

Tiny drops of water appear to cover sundews, a kind of plant. The "water" is actually a sticky substance that traps bugs drawn in by light bouncing off the water drops! This tricky adaptation helps the sundew catch the food it needs to be healthy.

Many species, or kinds, of plants use mimicry as a key to their survival.

On the Defense

Every animal and plant wants to keep from being a predator's next meal. Adaptations are sometimes used as a defense for plants. A defense is a way of guarding against an enemy.

Some plants have stems that are covered with sharp thorns. Predators don't want to get too close to those plants! Smelling bad and looking sick are other plant defenses that keep predators away. One kind of plant, called the sensitive plant, closes its leaves when it's touched. This startles animals that might try to eat it.

Sensitive means able to sense or feel changes in surroundings. Since plants can't move away from dangers, they must be ready to use other defenses to stay safe.

Mimicry or Camouflage?

Mimicry is often grouped with camouflage, another kind of defense. Camouflage is the colors and shapes on animals and plants that allow them to blend in with their surroundings.

Mimicry and camouflage can seem similar. In both cases, the plants appear to be something they are not. The main difference is that camouflaged plants don't want to be seen at all. They're great at staying out of sight! Plants that mimic are seen by predators, but they **resemble** something different than what they really are.

Super Fact!

Herbivores are creatures that eat only plants. All kinds of animals can be herbivores, including tiny bugs, rabbits, and elephants. Plants use camouflage and mimicry to protect, or guard, themselves from these creatures.

Living stones are a kind of plant that uses mimicry to stay hidden in plain sight. Although they are living plants, they look like pebbles or stones!

The Mate Mimic

Bees and flowers work together to survive. Bees visit flowers in search of **nectar**. At the same time, bees are helping with **pollination**. A flower's bright color and sweet smell are usually enough to draw in bees. But the bee orchid plant actually looks like a bee!

Bee orchids have pink petals and a yellow and brown center. From far away, the yellow and brown center looks exactly like a bee. Male bees are drawn to the orchid, thinking they see a female bee they can **mate** with.

Super Fact!

Bee orchids also *smell* like a female bee! The shape and smell of this orchid attracts, or draws in, male bees. Male bees then land on the orchid, picking up pollen and carrying it to the next flower.

Bee orchids are found in grassland habitats and their population is most stable, or steady, in mainland Great Britain.

Fly Fools

The fly orchid also uses sneaky mimicry. The plant's flowers have reddish-brown markings and are shaped like a fly's body. They attract other bugs looking for mates, like digger wasps. The flowers even smell like female wasps! Some bugs are fooled and land on the plant. The bugs leave without a partner, but they carry pollen with them.

Scientists think only a small number of fly orchids are pollinated this way, but the mimicry does help. Pollinated fly orchids produce many dust-like seeds that can spread widely.

Super Fact!

The fly orchid uses mimicry to draw in two species of digger wasps. These wasps help pollinate the plant.

The fly orchid is often hard to find in the wild, but the plant usually grows in groups of 10 or more.
BOOM!

Multitalented Mimic

In the rain forests of Chile and Argentina grows the *Boquila trifoliolata* vine. This vine climbs up trees for support. It is also a multitalented mimic. The vine doesn't just mimic one plant—it mimics many!

The vine changes the shape, size, color, and thickness of its leaves to match the plant it's growing on. If the vine's support tree has long, thin leaves, the vine will also have long, thin leaves. If the same vine crosses to a different tree, the vine's leaves change to match the new tree.

Super Fact!

The *Boquila* vine actually has two defenses: mimicry and climbing. Climbing keeps it off the ground, away from predators like weevils and leaf beetles.

Scientists think the *Boquila* vine changes its leaves to keep from getting eaten.

Smelly Scheme

While most flowers smell sweet to draw in pollinators, others use a stinky **scheme** to get bugs to visit them. The skunk cabbage is one of those smelly plants. It can mimic the smell of rotting flesh!

The skunk cabbage is a wildflower that grows in wet areas, like **swamps**. The awful smell attracts bugs, such as flies and gnats, that feed on **decaying** matter. When the bugs visit the flower, they pick up the flower's pollen. When the bugs move on, they become pollinators.

Super Fact!

The skunk cabbage is one of the only plants that can produce heat. This allows the skunk cabbage to grow even if the ground is frozen. It melts the snow around it!

Animals usually stay far away from skunk cabbage. When an animal eats the plant, it causes their stomach to feel like it's burning.

Deceptive Tricksters

A towering plant, the corpse flower, is world-famous for its brief blooming period—and powerful odor. Like the skunk cabbage, the corpse flower can mimic the smell of rotting flesh to attract pollinators. The plant's strong smell can draw in carrion beetles and flies from far away.

Plants that mimic are **deceptive** tricksters. They use colors, smells, and certain shapes to fool others. The next time you're in nature, look twice at your surroundings. Is there a plant trying to trick your eyes?

Super Fact!

The bloom of a corpse flower is a big event. The flower only blooms about once every 4 to 10 years. Then, once the flower is blooming, it only blooms for about 24 to 48 hours. People come from all over to watch it!

A corpse flower can grow more than 10 feet (3 m) tall.

GLOSSARY

decay: To slowly break down by natural processes.

deceptive: Meant to make someone or something believe something that is not true.

disguise: The state of having a false appearance.

habitat: The natural place where an animal or plant lives.

mate: To come together to make babies; one of two animals that come together to produce babies.

nectar: A sweet liquid made by flowering plants.

pollination: The transfer of pollen from one plant to another that can result in fruits and vegetables.

resemble: To look like something else.

scheme: A clever and often not honest plan to get something.

sneaky: Doing something in a quiet and unfair way.

swamp: An area with trees that is covered with water at least part of the time.

trait: A feature that differentiates one thing from another.

FOR MORE INFORMATION

BOOKS

Kaner, Etta. *Pretty Tricky: The Sneaky Ways Plants Survive.* Berkeley, CA: Owlkids Books, 2020.

Leedy, Loreen and Andrew Schuerger. *Amazing Plant Powers: How Plants Fly, Fight, Hide, Hunt, & Change the World.* New York, NY: Holiday House, 2018.

WEBSITES

Deceptive Colorization
nhpbs.org/wild/mimicry.asp
Discover different types of camouflage and mimicry here.

Mimicry
kids.britannica.com/kids/article/mimicry/353467
Learn more about mimicry in plants and animals at this website.

Publisher's note to educators and parents: Our editors have carefully reviewed these websites to ensure that they are suitable for students. Many websites change frequently, however, and we cannot guarantee that a site's future contents will continue to meet our high standards of quality and educational value. Be advised that students should be closely supervised whenever they access the internet.

INDEX